SAVE
THE PLANET
IN YOUR SPARE TIME

A CLIMATE PROTECTION HANDBOOK FOR THE BUSY PERSON

JOYCE MERCADO

SAVE THE PLANET
IN YOUR SPARE TIME

A CLIMATE PROTECTION HANDBOOK FOR THE BUSY PERSON

Praise for *Save the Planet in Your Spare Time*

As Mayor of a city with ambitious climate goals, I know that government needs the support of passionate and committed residents to achieve our goals. This book provides an excellent path for any concerned resident to follow.
—**Marilyn Ezzy Ashcraft, Mayor of the City of Alameda, California**

Many of us worry about climate change, but the good news is that there are so many small ways we can all contribute to addressing this very big issue. This book will give you the motivation and ideas you need to get started on your climate journey and be a part of the solution.
—**Danielle Mieler, Sustainability and Resilience Manager, City of Alameda**

In this inspiring, easy-to-read handbook Joyce Mercado takes us through her journey to becoming a climate activist, from small things she started doing around her home to writing articles and giving presentations. She has now started volunteering with Citizens' Climate Lobby to expand her influence even further to working for climate solutions with Congress. There's something here for everyone who wants to help to address climate change. Most of us feel overwhelmed by the immensity of the crisis and this can keep us from acting. The good news is that taking action is the way to not only make a difference, but to also make us feel empowered and give us the courage to continue.
—**Trish Clifford - Citizens' Climate Lobby volunteer - Liaison for California District CA08**

I have read Joyce's lovely handbook and really enjoyed her "Lewis and Clark" expedition into living a sustainable, earth-friendly life. Very well written, easy to follow, and a source of inspiration for even those of us who have been on the front lines of climate change activism for many years! Thank you!
—**Leslie Wharton, Chair, Elders Climate Action**

Save the Planet Press
Alameda, California 94501

jlmercado246@gmail.com

Editors: Lauren Davis and David Mercado
Cover Design: Kevin Mercado

Ordering Information:

Quantity sales. Special discounts are available on quantity purchases by corporations, associations, and others. For details, contact the "Special Sales Department" at the address above.

Save the Planet in Your Spare Time: A Climate Protection Handbook for the Busy Person/ Joyce Mercardo. -- 1st ed.

ISBN 979-8-218-31768-3–paperback
ebook edition also available
Library of Congress Control Number 2024909097

Acknowledgments

I would like to acknowledge my daughter Lauren Davis and husband David Mercado for editing this book, my son Kevin Mercado for the cover artwork, my brother Leonard Stanton for the photography, my friend David Teeters for his creative input, and my friend Terry Winckler for his publishing guidance.

Chapter 1

You Can Make a Difference

Are you wondering what difference one person can make to combat climate change? If so, this handbook is for you! The enormous challenge of climate change can be overwhelming, but a critical part of the solution is regular people like you pulling together to make a collective difference. To avert the most extreme impacts on the planet, the Intergovernmental Panel on Climate Change's 2023 report indicates we need to cut emissions in half by 2030 and bring them down to zero by 2050 to limit the temperature change to 1.5 degrees Celsius above pre-industrial levels. But what can one person do to help beyond reducing their own emissions? The answer is a lot!

I will share with you my journey of how a busy working mom of two became a friendly environmental activist to help save the planet in my spare time. What got me started? I read a *National Geographic* article on climate change. One chart showed the strong correlation between the earth's temperature and carbon dioxide concentration in the atmosphere over millions of years, then showed the current levels of carbon dioxide way above any time in the last 4 million years. This chart was my wake-up call. I realized I needed to do much more than change a few lightbulbs and recycle. I decided right then to make a difference in my community with a series of actions to not only change my habits, but to influence others to change theirs. Climate change can be downright depressing and overwhelming, but I found the cure for these feelings is taking personal action to make a difference. You can too! It feels great with each little win you achieve. Just do what you can fit into your lifestyle, taking it one step at a time.

I started by systematically reducing *my* carbon footprint (the amount of carbon dioxide and other carbon compound gases emitted because of an individual's actions). This action proved valuable beyond my own emissions reduction. I led by example to influence others and learned how to help others reduce their carbon footprints. Reducing your own carbon footprint is a great place to start. Once you get going, you can have a greater impact by taking the actions outlined in the remaining chapters of this handbook.

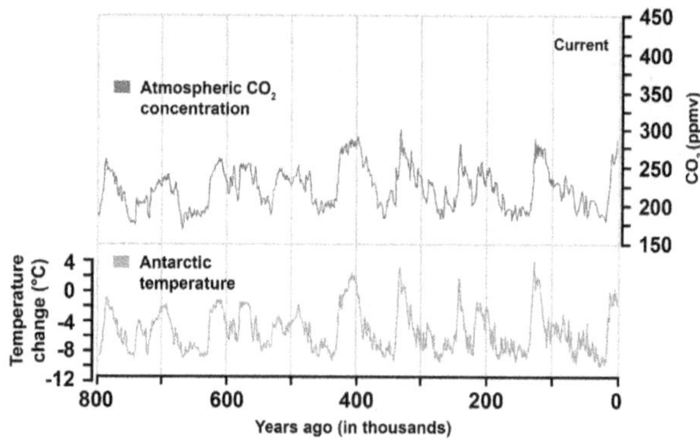

Correlation chart between temperature and carbon dioxide concentration in the atmosphere from The Royal Society.

I started reading up on ways to reduce my greenhouse gas emissions. Greenhouse gases are gases that absorb infrared radiation in the atmosphere, thereby warming the planet. There is a paper online called "Understanding and Responding to Climate Change, Highlights of National Academies Reports," which does a good job of explaining climate change in layperson's terms. It describes the various greenhouse gases and their sources, explains the effects of climate change, and focuses on the importance of replacing fossil fuels with new energy sources. The other item I found very helpful is the book *Drawdown: The Most Comprehensive Plan Ever Proposed to Reverse Global*

Warming edited by Paul Hawken. It goes through over 200 solutions to climate change. I have organized the carbon footprint reduction actions by emission categories. The category emission percentages listed for the United States are from the Environmental Protection Agency, while the emissions listed globally are from *Drawdown*.

Transportation
27% of emissions in the United States
14% of emissions globally

The first thing I did to reduce my carbon footprint was to dust off my bike in the garage and start using it to run errands around town instead of driving my car. I then added a basket, rack, and panniers (baskets that attach to the bike rack in the back) to my bike, and now I can ride instead of driving to go grocery shopping, to the post office to mail packages, etc. A bungee cord attached to the bike rack proved useful for hauling large items like a giant package of toilet paper. Next, I added front and rear lights so I could ride in the evening as well. I saved a lot on gasoline and wear and tear on the car. It was also nice not having to hunt for a parking space. Any pole would do. I eliminated having to feed the parking meter, and biking is excellent exercise as well.

I was diligent about carpooling for kid activities such as sports practices and games. I was taking my daughter to and from gymnastics practice four days a week until another mom organized a carpool for three of the gymnasts, which cut our trips by two-thirds! The unexpected benefit of carpooling kids to events was getting to hear the teenagers talking to each other about what was going on in their lives as if I wasn't there—much better than the single-word responses so common among teenagers. Then I learned about the UPS trick of organizing your route in a clockwise manner to eliminate left-hand turns, which leads to less idling at intersections. Before making that switch, I had been taking my daughter to gymnastics in a counterclockwise manner for years. The carpooling and UPS trick saved gas and time, and I learned more about what was going on in my kids' lives.

To save further on gasoline, I clean out the back of the vehicle periodically of excess stuff to lighten the load, check the tire air pressure once a month, consolidate trips, accelerate smoothly, and let my foot off the gas pedal much earlier to coast to a stop.

I also had the opportunity to work remotely, which saved commuter trips to work. If you cannot do this 100% of the time, even once or twice a week helps a lot.

We then replaced one of our gas-powered vehicles with an electric vehicle, taking advantage of the Inflation Reduction Act tax incentives. It sure is nice to avoid gas pump prices and fuel our electric vehicle with our city's 100% renewable energy. But does it make sense to buy an electric vehicle where you live that may not have 100% renewable energy? According to a United States Environmental Protection Agency report, "FACT: Electric vehicles typically have a smaller carbon footprint than gasoline cars, even when accounting for the electricity used for charging."

Electricity

25% of emissions both in the United States and globally

Buildings

13% of emissions in the United States and 6% globally

We installed solar panels on our roof taking advantage of available rebates. This action dropped our electricity bill to almost nothing each month. I recognize not everyone can buy solar panels. One thing that can help instead, is some power companies offer an option of renewable power such as wind, thermal and solar, typically at a slightly higher rate. We took that option until we installed our solar panels. Some locations in the country allow you to select which power company to use, so you can look for one that focuses on renewable power.

I got a home energy efficiency audit from the local power company and implemented many of their suggestions. Most power companies provide such a service for free or for a small fee.

I replaced the remaining lightbulbs with compact fluorescent lightbulbs (later those were replaced by LEDs). I found bulbs that looked great in our antique light fixtures with a little searching. I installed lined curtains in the sunroom for winter months, and wood shutters in the front room for heat retention in the winter and to keep it cooler in the summer. I replaced the worn-out weather stripping on the back door and windows to prevent heat loss. There is a clear insulating film that can be applied to windows as well. We also had a mail slot by the front door that was leaking air, so I made a felt cover flap to help stop the heat loss. The dog kept attacking the mail as it came through the slot, so we eventually filled in the slot and installed an exterior mailbox. Implementing those energy audit recommendations reduced our electric and gas bills.

In winter, we lowered the thermostat to 65 degrees instead of 68 degrees and put on a sweater. We do not have air conditioning as it rarely gets too hot in our town, but if you have air conditioning, for days that are not too hot, use a fan instead and save electricity. Having a technician maintain your heater and air conditioner is also important for them to run efficiently and last longer.

My son and I put up a clothesline. I had put off this action, as I was not sure how to do it. It turned out to be a very simple process involving two screws at one end and a hook for the other. We had it done in ten minutes, and that includes dropping the screws in the dirt a few times. I would do a load of laundry the night before, put it up on the line after getting the kids off to school in the morning, then the kids would take it down and fold it after school. All you need is sunshine and/or a breeze for line drying to work. The only thing that did not work well were the towels. They would be stiff as boards. To fix this, I learned to fluff them in the dryer for a few minutes before hanging them on the line.

I also switched to washing clothes in cold water when I learned that up to 90% of the energy it takes to do a load is for heating the water. Today's detergents are formulated to wash in cold water. It works just fine. Also, your clothes last longer with a cold-water wash and colors stay brighter, as cold water is gentler on your clothes than hot water. I was also diligent about doing full loads of laundry, so I got to do laundry less often. I also switched to detergent strips (thin small sheets of concentrated detergent) that come in paper packaging instead of liquid laundry detergent, which comes in large plastic jugs. Liquid laundry detergent is mostly water and takes more energy to ship than laundry strips. Plus, the plastic jug is more material to recycle than the small box or paper envelopes the laundry strips are packaged in.

We upgraded our electrical panel so we could install a charger when we purchased an electric vehicle, and we replaced our gas dryer with an electric one for the days a clothesline will not work. Migrating from gas-powered to electric-powered appliances reduces emissions. We took advantage of rebates and tax credits from our local power company and the Federal government. Be sure to follow all the requirements for the rebates, looking them up in advance of your replacement purchases.

We explored replacing our gas-powered furnace with a more efficient electric-powered heat pump to reduce emissions. Heat pumps work like a refrigerator, but in reverse, taking warmth from the air and creating heat to pump throughout the home. Unfortunately, the ducting system hidden behind our ceilings was too narrow for the more powerful heat pump furnace. Hopefully, you will have better luck with that option. We plan to replace our gas-fired hot water heater with a more efficient electric-powered hot water heater the next time it needs to be replaced.

Industry
24% of emissions in the United States
21% globally

Although we may not run industries, we can help by reducing the things we consume. Everything we consume takes energy—getting the raw materials, producing, shipping, and eventually disposing of the finished items, with greenhouse gases being emitted at each step. I was already doing a good job of recycling and composting, which reduces methane production in landfills, but I needed to reduce my consumption of single-use items.I made a shopping list and picked up reusable items, including a reusable coffee mug instead of using all those paper cups, sleeves, and lids (most coffee shops give me a discount with my reusable mug); a reusable water bottle (use a filter if necessary) to eliminate purchasing plastic water bottles, which take a lot of energy to ship. We also saved trees by switching to cloth napkins and dish towels instead of paper napkins and paper towels, and we purchased reusable shopping bags, including a cloth bag that rolls up into a small bundle to keep in your purse to have handy for non-grocery shopping.

We are also saving trees by removing myself from catalog company mailing lists, reducing the amount of junk mail I receive, and we switched from print newspapers to online newspapers. I got into Freecycle (sometimes called Buy Nothing or Trash Nothing depending upon where you live) which allows folks to give away or get unwanted items for free. I repaired things instead of replacing them, such as mending clothes. I also mend clothes for my neighbor. I decline straws at restaurants and plastic utensils and chopsticks when getting to-go food. We bought glass and metal straws to use at home for cocktails. We switched to paperless billing. My daughter switched to reusable period pads that can be washed and reused after each use. Many of these changes also saved money besides reducing greenhouse gas emissions.

Agriculture
11% of emissions in the US
24% of emissions globally

I learned about the negative impact of cows and sheep on the planet. They burp up methane gas, which is a powerful greenhouse gas. Furthermore, the land used for grazing livestock and growing their feed accounts for almost 80% of farmland, yet produces less than 20% of the world's supply of calories. Forests are being cut down for cattle grazing and to grow feed for them. So, I severely reduced beef and lamb from our family's diet. This was not as difficult to do as I thought. Instead of hamburger, I used ground turkey to make chili and meatloaf, and I made tacos with shredded chicken. I tackled dairy as well since it comes from cows, changing my standard breakfast from yogurt and fruit to oatmeal with fruit or a smoothie made from veggies and fruit. Switching to soy or almond milk is also helpful. Then when I learned that even chicken and turkey cause more greenhouse gas emissions than beans, I started substituting beans for those meats as well. I learned to cook some vegetarian dairy-free dishes. I posted pictures and recipes of vegetarian dishes on Facebook. This way friends could try out new easy-to-make vegetarian dishes with very little effort on my part.

Food waste is a big problem as well, with forty percent of food wasted in the United States. I became more diligent about only buying what I needed for the week at the grocery store. When vegetables become wilted, I make a stir fry or soup or put them in a smoothie. The freezer has become my friend for leftovers. I also buy an imperfect produce box of fruits and vegetables most weeks, which uses produce that would otherwise go to waste. My change to a primarily plant-based diet led to my cholesterol going down!

I also learned that palm oil leads to the destruction of rainforests by bulldozing or burning in Southeast Asia, Africa, and Latin America in order to plant palm tree plantations. Palm oil is not good for you either, containing a lot of saturated fat. I started looking at ingredients in packaged foods and eliminated those with palm oil. Bye-bye microwave popcorn. I instead put kernels in a microwavable container with a lid and make popcorn without all the additives. Watch out for crackers and cookies as well.

Reverse Carbon Dioxide Emissions

We planted fruit trees in our yard, which is a real treat! We enjoy figs, pomegranates, apples, persimmons, and mandarin oranges. The fruit trees take carbon dioxide out of the air and produce oxygen in addition to feeding us. We also planted a vegetable garden with tomatoes. After we had no more room for trees, I started a monthly donation to The Nature Conservatory, which buys land to preserve nature and reforest areas with new trees. My young daughter was passionate about the rainforest, so as an incentive for her to do her reading (she has dyslexia), we purchased an acre of rainforest at the Nature Conservancy when she hit certain milestones. We also replaced our lawn with drought-tolerant plants. No more mowing grass and blowing bits with gasoline-powered leaf blowers.

In summary, here is a list of the carbon footprint reduction items I took:

- Ride a bike for errands.
- Carpool.
- Arrange errands in a clockwise manner.
- Clean out trunk/back end of vehicle.
- Check tire pressure monthly.
- Consolidate trips.
- Accelerate smoothly and let my foot off the gas pedal earlier when coming to a stop.
- Work remotely.
- Replace gas-powered vehicle with electric vehicle.
- Replace gas-powered dryer with electric dryer.
- Install solar panels.

- Get a home energy efficiency audit.
- Replace incandescent light bulbs with LED light bulbs.
- Install thick curtains in the sunroom for winter.
- Install clear insulting film on windows.
- Lower the thermostat to 65 degrees in winter.
- Maintain heater (and air conditioner if you have one).
- Install clothesline.
- Wash in cold water.
- Use laundry strip detergent.
- Compost food scraps and food-soiled paper.
- Use a reusable coffee mug.
- Use a reusable water bottle filled with tap water.
- Use cloth napkins and dishcloths instead of paper napkins and paper towels.
- Use reusable shopping bags.
- Leverage Freecycle to get rid of stuff or get stuff used.
- Repair things instead of replacing them.
- Decline single-use items when getting takeout.
- Switch to paperless billing.
- Switch to reusable period products.
- Eliminate beef and lamb from diet.
- Switch to a plant-based diet.
- Purchase imperfect produce.
- Avoid products with palm oil.
- Avoid food waste.
- Plant trees—lots of trees.
- Replace lawns with native plants and vegetable garden.

I hope you got a few new ideas for reducing your personal greenhouse gas emissions in this chapter. Now let us move on to how to broaden your impact beyond your own carbon footprint.

Chapter 2

The Climate Protection Checklist and Resource List

While I was reducing my carbon footprint, I wanted to do more to influence others to reduce theirs. A critical part of protecting the climate is getting more people to change habits and complete green projects, which is not always easy. It helps to be very specific, make things as easy as possible, and incentivize people with reasons they should act. To address these challenges, I created a climate protection and resource list to distribute to others. The first page of the list has about 70 actions you can take to reduce greenhouse gas emissions and the second page lists various local resources to help you tackle the list. I mention at the top of the checklist that besides helping protect the climate, many of the actions also save money. The idea is to post the document in a prominent location at home and check things off as you complete projects and change habits.

You can find the latest version of my checklist on the website **casa-alameda.org/resources.** This project took quite a bit of time. I drafted the checklist while I was valet bike parking (watching over bikes while bicyclists enjoyed a community event), did extensive research on resources available to our town's residents, and then typed it up. I have been maintaining it over the years as I find new ideas for reducing emissions as well as new resources to help people on their emission reduction journey. With the head start I have given you with my version, it will not take you long to tailor it to your community. If you have less time, you can download and tailor the more generic checklist I created without the resource list on the same website.

Once I completed the checklist, the first thing I did was find a home for it online. Community Action for a Sustainable Alameda (CASA) posted it on their website for download by the public. I reached out to local community groups to distribute printed copies for free and kept them stocked with checklists. I have handed out hundreds over the years by giving one to everyone I met while I was out and about. For example, when I was at sporting events for my kids, I would give out checklists to fellow parents. At first, this felt a little intimidating, as I did not know how people were going to react, but the reception was overwhelmingly positive. I would just introduce myself as Kevin or Lauren's mom and say I have something I put together for them, and hand them the checklist. I would explain the concept behind it and people loved it, especially the resources available to help them. Most people I spoke to were interested in what more they could do to help the environment. I pointed out that many of the items in the checklist save the family money as reducing your carbon footprint primarily is about saving resources. Pointing this out helped motivate people to act on the checklist even if they were not as passionate about the environment as I am.

Each time I gave out the checklist I asked if the individual had a way of getting it out to more people. This approach got the checklist published in a homeowner's association newsletter. I also contacted all the non-profits listed on the resource page of the checklist and got many of them to distribute the checklist or post a link to it on their websites. The local historical society included it in their newsletter. I also got together with a friend at CASA and we handed them out at the Farmers Market and at our city's Earth Day celebrations. I also stocked the checklists at various venues in town such as cultural centers and marketplaces with their permission and handed them out at climate protection talks.

These combined actions distributed thousands of copies of the checklist to members of the Alameda community.

I am a member of the Golden Gate Bird Alliance (GGBA), a local bird advocacy non-profit organization. One of their newsletters asked for volunteers to be involved in a climate protection initiative. GGBA is extremely interested in protecting the climate as National Audubon's 2019 report shows that over two-thirds of North America's birds—389 species—are at risk if the current pace of climate change continues. Birds are affected in part because earlier springs bring out insects earlier, while the migration timing stays the same, so many birds miss out on the feast of insects. I signed up and was invited to take part in a GGBA meeting on climate change and how it affects birds. I told them about my checklist. They were interested, and I modified it to say GGBA at the top with their logo, changed the introductory paragraph to talk about birds, and made it a little more generic without the resources list so it could be used throughout the San Francisco Bay Area. GGBA leaders reviewed the checklist and suggested some changes which I incorporated, and they posted it on their website.

Getting my name out there with the checklist opened many doors for me. That is the next chapter.

Help Protect the Climate! Each of Us Can Make a Difference!

Post this checklist in a prominent location at home. Tackle an item or two at a time, checking them off as you complete projects and change habits. In addition to protecting the climate, you will save money and incorporate a little more activity into your lifestyle. Leverage the plethora of resources on page 2 for assistance completing the checklist items.

Transportation - Every gallon of gasoline burned creates 19.4 lbs of CO_2 (from US EPA)

The most critical change needed in California is to drive less!

- ☐ Bike or walk
- ☐ Take public transit (relax - no fighting traffic)
- ☐ Carpool for work, kid activities, meetings, events, clubs, etc.
- ☐ Work from home if possible (even 1 or 2 days/week helps)
- ☐ If you buy a car, go electric, smaller is better
- ☐ Consider going car-free (use carshare if need car)

Drive more efficiently:

- ☐ Clear out items from trunk/rear of vehicle (do seasonally)
- ☐ Check tire air pressure monthly
- ☐ Keep regular vehicle maintenance schedule
- ☐ Do not exceed speed limit; accelerate/brake smoothly
- ☐ Consolidate trips
- ☐ Organize route in clockwise pattern to minimize left-hand turns

Conserve Energy at Home

Heating/Cooling:

- ☐ Insulate attic ☐ Replace gas furnace with heat pump
- ☐ Replace/clean furnace/AC filters at least every other month when in use
- ☐ Caulk/weather strip windows & exterior doors
- ☐ Lower (raise) thermometer setting 2 or 3º F in winter (summer) – Use smart thermostat
- ☐ Use a fan to cool instead of A/C (turn off when leaving room)
- ☐ Install insulated curtains/blinds & plastic film on windows

Lighting:

- ☐ Replace incandescent with florescent or LED bulbs

Water Heater:

- ☐ Properly install tank & hot water pipe insulation
- ☐ Replace gas water heater with heat pump water heater

Appliances:

- ☐ Use appliance energy-saving options like air dry
- ☐ Purchase only Energy Star® appliances
- ☐ Consider replacing old fridge & doing w/o 2nd one
- ☐ Clean refrigerator coils every 6 months

Laundry:

- ☐ Wash full loads only
- ☐ Wash and rinse with cold water
- ☐ Clean dryer filter before each load
- ☐ Don't overheat clothes in dryer (use moisture sensor option)
- ☐ Even better, install and use a clothesline or rack

Dish Washing:

- ☐ Skip garbage disposal; scrape food bits in green bin
- ☐ Only run dishwasher when completely full
- ☐ Fill sink/basin rather than letting water run over dishes down drain

Miscellaneous:

- ☐ Turn off items when not in use (lights/TV/computers…)
- ☐ Take shorter showers
- ☐ Consider installing solar panels
- ☐ Microwave when possible (more efficient than stove/oven)
- ☐ Use smart power strips & unplug electronics when not in use
- ☐ Swap gas appliances for electric
- ☐ Install reduced flow shower/faucet heads

Reduce, Reuse, Recycle and Rot (Compost) – The Four R's

- ☐ Put paper, plastic, glass & metal in blue bin
- ☐ Put yard trimmings, food & food soiled paper in green bin
- ☐ Reuse mug for coffee/ tea instead of paper cups
- ☐ Reuse cloth bags instead of paper/plastic for all shopping
- ☐ Buy reusable bottles and use tap water (filtered if you like) instead of bottled water
- ☐ Sign up for Freecycle™ to give/get local free items
- ☐ Decline plastic straws/utensils & any extras you don't need

- ☐ Repair items instead of replace if salvageable
- ☐ Buy used or made-from-recycled materials products instead of new
- ☐ Use dishcloths/rags/mops instead of paper towels/other disposables
- ☐ Donate/sell gently worn, no longer wanted items
- ☐ Pack no-waste lunches (see back for help)
- ☐ Switch to paperless billing ☐ Reduce junk mail (see back for help)
- ☐ Use reusable period products and reusable diapers
- ☐ Reduce food waste (see back for help)

Other Ways to Help

- ☐ Plant trees (lots of trees); grow a fruit and vegetable garden
- ☐ Shop locally; buy locally made/grown goods
- ☐ Eat a plant-based diet; avoid beef, lamb, and palm oil
- ☐ Fly less (consider a staycation)

- ☐ Use rake, broom & push mowers for yard work
- ☐ Vote and campaign for environmentally minded officials
- ☐ Contact representatives at all levels about climate change
- ☐ Volunteer with/donate to organizations that protect the climate

For an electronic copy of this checklist visit our website at www.casa-alameda.org

Questions? Email Joyce Mercado at jlmercado246@gmail.com

April 2023

Helpful Resources Available for Alamedans!

Transportation

- **East Bay Bicycle Coalition:** Free bike safety classes. https://bikeeastbay.org/
- **AC Transit:** Trip Planner, Bus Schedules/Routes, real time arrival information using NextBus or Routeasy apps or www.actransit.org
- **BART:** Quick planner, real time departures, schedules https://www.bart.gov/
- **Ferry Service:** https://sanfranciscobayferry.com/
- **511.org:** Public transit trip planner, find a carpool, get FasTrak®, check traffic conditions, bike routes, etc. Call 511 or www.511.org
- **Golden Gate Electric Vehicle Association:** Nonprofit provides guidance on electric vehicles, rebates, charging, etc. www.ggeva.org
- **Drive Clean Bay Area:** Lots of information about Electric Vehicles. www.DriveCleanBayArea.org
- **Shared Mobility resources** (carshare) https://www.alamedaca.gov/RESIDENTS/Information-for-Residents/Getting-Around-Alameda#section-7

Energy and Water Conservation

- **PG&E:** Residential rebates & energy conservation tips www.pge.com/myhome/saveenergymoney
- **Alameda Municipal Power:** Residential and business rebates https://www.alamedamp.com/
-
-
- **East Bay Mud:** Water saving tips, gear & rebates http://www.ebmud.com/water-and-drought/conservation-and-rebates/

Reduce, Reuse, Recycle and Rot (Compost)

- **Alameda County Industries (ACI):** What goes in each bin, https://alamedacountyindustries.com/alameda/residential/resources/
- **StopWaste.org:** Great information including household hazardous waste disposal, bay friendly gardening, green building, environmental purchasing links, etc. www.stopwaste.org/home/index.asp
- **Universal Waste Management, Inc.:** Free recycling of electronics, textiles, non-useable clothing, pillows, curtains, bedding, linens, towels, small household appliances, paper products and more at 721 37th Ave. Oakland https://dtsc.ca.gov/hw-projects/universal-waste-management-inc/ 888-832-9839
- **Sell/Buy Used Items:** www.ebay.com or www.craigslist.org or Facebook Marketplace
- **Bay Area Salvage Yards:** www.ohmegasalvage.com/resources/bay-area-salvage-yards/
- **Habitat ReStore:** Accepts donations and resells new/gently used building materials and household goods. 9235 San Leandro Blvd, Oakland, CA 94603 http://www.habitat.org/restores
- **No-waste Lunches:** Information on how to pack waste free lunches https://www.epa.gov/students/pack-waste-free-lunch
- **Junk Mail Reduction:** fee service www.41pounds.org; free kit http://bayarearecycling.org/content/learn-how-reduce-your-junk-mail

Tree Planting, Gardening and Food

- **100K Trees for Humanity:** Tree planting organization https://www.100ktrees4humanity.com/
- **Alameda Backyard Growers:** Building community 1 veggie at a time. Volunteers provide monthly education, glean excess fruit from trees for donation to Alameda Food Bank, and work with City to plant fruit trees in Alameda. http://alamedabackyardgrowers.org/
- **The Nature Conservancy:** Protects natural lands, restores forests https://www.nature.org/en-us/
- **Reduce Food Waste:** Food waste reduction tips (food planning, storage & preparation) at StopFoodWaste.org and Savethefood.com

General Community Assistance

- **Climate Reality Project:** Training for speakers on climate protection https://www.climaterealityproject.org/
- **Climate Action Now:** An app for writing state and federal government representatives and corporations on climate protection
- **Vote.org:** Register to vote

Chapter 3

Anyone Can Be a Writer

A simple way to influence others to reduce their greenhouse gas emissions is to write letters to the editor of your local newspaper. It is an easy way to reach thousands of people, and elected officials monitor letters to the editor to keep a finger on the pulse of their community. I started doing this as I worked to reduce my carbon footprint. Look at your local newspaper's letter to the editor page; they usually provide information for submitting letters, a word limit, and any other requirements. The keys to writing an effective letter to the editor are to make it short and sweet, rather than a tome, and to share your tips and techniques around one topic at a time. Include what you learned, the benefits gained—especially if some benefits were unexpected—and a bit of humor if you can. Do not tackle too much in a single letter. The tone should be friendly, as if you are speaking to a friend. No ranting, preaching, or lecturing allowed. I have included a couple of letters to the editor here about my journey to reducing my carbon footprint as examples.

Give Pedaling a Try

With overwhelming evidence of climate change, I thought I had better start doing more than just the little things like turning off lights. Since I could stand to lose a few pounds (OK, several pounds) I decided to try running errands on my bike instead of driving.

The keys to starting were: 1) getting a bike lock and a basket that hooks onto the bike for carrying purchased items, 2) getting a bike safety check and tune-up (did this at Alameda Bicycle on Park Street—wonderful folks there), and 3) keeping the bike, helmet, and lock in the front entryway so there is no hassle grabbing it to head out quickly.

It worked. It is a lot easier than I thought it would be. I have been running most errands around town on my bike since the beginning of the year, and I am no athlete. Fortunately, Alameda is flat.

Since I do not have to hunt for parking (any pole will do) it does not take that much longer than driving. No feeding parking meters either. I have been surprised at how much gas I am saving. Since I am fortunate enough to telecommute almost every day, I only fill up the tank once every six weeks or so.

The exercise is making me feel better too. If a forty-something, non-athletic, working mom of two active kids can do it, so can you. I hope to see many more fellow Alamedans on your bikes around town. Give it a try. You will be glad you did.

Joyce Mercado

Carpooling

According to an International Council for Local Environmental Initiatives greenhouse gas emissions study of Alameda, 53% of our emissions come from transportation. One of the most important things we can do locally to protect the climate would be to drive less.

The cost savings are significant: less money for gas, deferred vehicle maintenance, and less mileage on your car, which postpones your next vehicle purchase. The basic options are public transit (relax or get stuff done while commuting and avoid parking hassles), biking or walking (most of us could use more exercise), and carpooling. I think that the last option is often overlooked.

For years I drove my daughter to Bay Island Gymnastics after school for workouts four times per week until another mom organized a carpool for three of the gymnasts. I asked myself why I had not thought of this years ago? Now, instead of three parents having to schedule time away from work to take our girls separately to and from the gym four times per week, we share the driving and take two-thirds fewer trips each. We have bought back so much time as parents. The girls love it, we are saving money, and it helps the environment.

Wouldn't it be great if everyone in Alameda organized kid carpools for sports practices, games, scout meetings, etc. when biking or walking is not practical? Consider carpooling.

Joyce Mercado

I had a thought that it would be helpful if someone wrote weekly climate protection tips for our local newspaper to help people change their habits. Then one day it occurred to me that that person could be me! So, I wrote a couple of sample tips and sent them to the editor of our local paper. He said yes and off I went. You could do the same thing. Each tip did not take long to write, yet reached thousands of people each week—a very effective use of time. After a year or so, I received funding from a local non-profit called Community Action for a Sustainable Alameda (CASA) to continue my green tips when the newspaper started to charge for publishing them. Here are some samples to get you started.

Fill Up at The Pump Less Often—Drive Efficiently

Hauling around a lot of excess stuff in your trunk? Clean it out except for emergency gear. It takes more gas to haul around excess weight. Check tire pressure once a month and improve gas mileage by up to three percent. Try taking your foot off the gas a little sooner as you approach stops and let momentum work for you. Then use a light touch on the gas pedal to get going again and do not exceed the speed limit.

When was the last time you had your car tuned up? Check your vehicle's regular maintenance schedule to see if it is time again to take it in. Do you have two drivers and vehicles at home? Is the driver who drives the most using your most fuel-efficient vehicle? If not, consider switching. Arrange your routes in a clockwise pattern to minimize left turns (UPS trick).

Washing Clothes

Wait until you have a full load before running the washer. Running the washer less often saves electricity, and besides you have got better things to do than laundry. Use cold water to reduce natural gas consumed by heating the water.

Drink Tap Water Instead of Bottled Water

Save money and the environment by purchasing a refillable bottle and using tap water instead of bottled water. It is far more efficient to reuse a single refillable bottle hundreds of times than to recycle hundreds of empty plastic water bottles. Even more importantly, bottled water is heavy and takes lots of fuel to ship from bottling plants to stores, and finally to your home. Just turn on the tap instead, avoid all that shipping fuel waste, and enjoy fine-tasting East Bay MUD water. If you are concerned about your pipes, consider using a water filter.

Be ready to say yes to unexpected developments! My letter to the editor about putting up a clothesline led to an inquiry from the editor to see if I was interested in writing a monthly environmental column when another columnist left. I immediately said yes even though I was not entirely sure what I was going to write. I got ideas from friends and wrote columns like the ones below. An interesting development was when a Japanese English language testing outfit inquired if they could use one of my columns for an English language test, and of course, I said yes.

For the busy people out there, each column can take just an hour or two to write and was an effective way to reach thousands of people in our community. Here are some examples of columns I have written over the years. The newspaper editor can provide you with maximum word count. The tips for writing columns are the same as the ones for writing letters to the editor. I always include actions people can take to help protect the climate. I get ideas for columns from newspaper articles I have read, the Sustainability Manager in our city, the local power company and garbage company websites, and fellow members of our local climate protection non-profits.

Alameda Sun Article—The Time for Climate Protection Action Is Now

By Joyce Mercado
Community Action for a Sustainable Alameda

The Intergovernmental Panel on Climate Change (IPCC) issued its March 2023 report, and it is alarming. What is the IPCC? According to the

IPCC's Fact Sheet, "The Intergovernmental Panel on Climate Change (IPCC) is the international body for assessing the science related to climate change. The IPCC was set up in 1988 by the World Meteorological Organization (WMO) and the United Nations Environment Programme (UNEP) to provide policymakers with regular assessments of the scientific basis of climate change, its impacts and future risks, and options for adaptation and mitigation." It has 195 member countries and hundreds of scientists. The 2023 IPCC report warns that to limit warming to 1.5 degrees Celsius to avert the most severe consequences, we must immediately make deep cuts to greenhouse gas emissions across all sectors this decade, get to zero emissions by the early 2050s, and then begin to reduce existing greenhouse gases in the atmosphere. We have no more time to waste. The good news is we have solutions at hand and there is plenty each of us can do to make a difference.

There is a lot more one can do besides reduce your own greenhouse gas emissions, but that is a good place to start. Green your ride by bicycling, walking, carpooling, and making your next vehicle purchase an electric vehicle. Reduce emissions at home by insulating, changing lightbulbs to LEDs, and converting gas appliances to electric, like replacing your gas furnace with a heat pump and gas water heater with a heat pump water heater. Reduce meat in your diet, especially beef and lamb, move towards a plant-based diet, and tackle food waste at home. Plant some trees. For a free sapling or two, visit the Rotary booth at the Spring Shindig Saturday, April 15th, 12:00-3:00 pm at the Alameda Point Gym and Field (1101 W Red Line Ave.) For more ideas, see the CASA climate protection and resource list here: **casa-alameda.org/wp-content/ uploads/2022/07/CASA-Climate-Protection-Checklist-July-2022.pdf**.

Then you can go beyond reducing your own carbon footprint by sharing your experience with greenhouse gas reduction actions with others. Talk to your friends and neighbors. Point out the CASA Checklist and Resource List to them as well. Write a letter to the editor about one climate protection change you made, tips to implement, and the benefits received. Reach out to the Mayor and City Council to let them know taking action to protect the climate is important to you (you can get their contact information at **alamedaca.gov/GOVERNMENT/ Elected-Officials**). There is a free iPhone app called Climate Action Now, for reaching out to state and federal officials as well as corporations to encourage them to take specific actions to protect the climate. I spend five minutes every morning sending out about ten scripted emails. It is easy, effective, and fast to use. Campaign for, donate to, and vote for politicians who will act to protect the climate. Contribute to and/or volunteer with local non-profits that work to protect the climate like Bike Walk Alameda, Alameda Backyard Growers, and Community Action for a Sustainable Alameda. If you have a group interested in climate change, request a CASA presentation either in person or via zoom, by emailing **info@casa-alameda.org**.

Individuals can make a difference and it is going to take all of us working together, taking it to the next level, to meet this challenge and limit warming to 1.5 degrees Celsius. Let's not wait. Act now!

Alameda Sun Article—Save Money on Your Heating Bills this Winter

By Joyce Mercado
Community Action for a Sustainable Alameda

Now is the time to prepare your home for the upcoming cooler temperatures to reduce natural gas and electricity usage over the fall and winter. The consumption of natural gas in buildings contributes 27% of Alameda's greenhouse gas emissions. Weatherizing your home will not only protect the climate, but it will also reduce your energy bills. So where should one start?

If you have not insulated your attic yet, that is the most effective thing you can do to reduce your heating bills. Insulating materials' resistance to conductive heat flow is rated in terms of its thermal resistance or R-value. The higher the R-value, the greater the insulating effectiveness. Insulation with at least an R-30 rating is recommended in the attic in the San Francisco Bay Area.

Next, consider getting a smart thermostat. What makes a thermostat smart? They are programmable to create a heating schedule; one temperature during the day, and a lower temperature at night, for example. Some smart thermostats can sense if anyone is at home, and if not, lower the temperature. App alert options can remind you when to change heating filters. Energy reports can tell you how much energy you are using and how to cut back. Alameda Municipal Power offers rebates on smart thermostats. See **alamedamp.com/407/Rebates-and-Incentives** for details.

How long has it been since you installed

weather stripping around external doors and older windows? Weatherstripping closes gaps that let in cold air and allow heat to escape but can deteriorate over time. Check your weather stripping and install new stripping as needed. It's a low-cost technique.

Caulking is also cheap and effective. Caulk around windowpanes with leaks. Also caulk around where plumbing, electrical wiring, or ducting enters through walls to seal those gaps. For larger gaps, use foam sealant. Use caution when sealing leaks around fireplace chimneys, furnaces, and water heater vents. These call for fire-resistant materials such as sheetrock, sheet metal, and furnace cement caulk.

Other measures include using heavy curtains or shades on windows to retain heat, closing the fireplace damper when not in use, changing furnace filters regularly, installing foam gaskets behind electrical outlets, and putting on a sweater and slippers so you can lower the heat a bit.

Let's all lower our heating power consumption this season, protect the climate, and reduce our energy bills.

Alameda Sun Article—Plant-Based Diet Benefits

By Joyce Mercado,
Community Action for a Sustainable Alameda

Food production is a major contributor to climate change, creating a quarter of greenhouse gas emissions globally. Food that comes from animals makes up two-thirds of all agricultural greenhouse gas emissions and uses more than three-quarters of all agricultural land (to raise the animals and to grow to feed the animals). If everyone adopted a plant-based diet instead of our Western meat-centric diet, we could cut food-created greenhouse gas emissions by as much as 70 percent (vegan diet)

or 63 percent (vegetarian diet, which includes dairy and eggs).

In terms of land use, it is much more efficient to grow plants than to raise animals and the food to feed them. Plant-based diets can feed fourteen times the number of people on the same amount of land as meat-based diets. If more people converted to a plant-based diet, land previously used to raise animals and their feed could be reforested, creating much-needed carbon sequestration.

Given droughts around the world with our growing temperature increases, saving water has taken on more urgency. Producing animal products uses far more water than producing plant foods. It takes 4000 liters of water to produce a kilo of beef, while it only takes 1071 liters of water to produce a kilo of legumes.

There are health benefits to a plant-rich diet as well. According to the Mayo Clinic, "A plant-based diet can reduce your risk of heart disease, diabetes, high blood pressure, obesity and certain types of cancer." A typical plant-based diet should consist of half your plate of leafy greens, one-quarter of your plate of legumes, and the other quarter of your plate of grains. It is important to include healthy fats as well such as avocado, olive oil, nuts, and seeds. Refer to this Mayo Clinic article on getting the proper nutrition on a plant-rich diet: **mayoclinic. org/healthy-lifestyle/nutrition-and-healthy-eating/in-depth/vegetarian-diet/art-20046446.**

A good place to start your plant-rich diet journey is to cut back on beef, dairy, and lamb. Cattle and sheep burp up methane gas (a greenhouse gas that is 86 times more powerful than carbon dioxide over a 20-year period). They also cause deforestation to expand grazing pastures, which releases a large amount of carbon dioxide as well as eliminates a carbon sink. The raising of cattle has caused the destruction of 80% of the Brazilian rainforest. Cattle and growing feed for cattle is the number one cause of deforestation globally. According to Project Drawdown, "if cattle were their own nation, they would be the world's third-largest emitter of greenhouse gases."

One does not have to go to a 100% plant-rich diet overnight to make a difference. On my journey towards a plant-rich diet, I first cut back on beef, lamb, and dairy by substituting ground turkey for ground beef in recipes, switching to oatmeal instead of yogurt for breakfast, and eating more chicken, eggs, and fish. Then I slowly cut back on poultry and fish and added legumes and more whole grains like quinoa, farro, and barley to my diet. Good books to reference include Vegan Reset: The 28-Day Plan to Kickstart Your Healthy Lifestyle and 28-Day Plant-Powered Health Reboot. Some great places to buy plant-based foods in Alameda are the Grocery Outlet and the Alameda Marketplace. Many of our restaurants in town offer vegetarian or vegan dish options. The Butcher's Son in Berkeley is a vegan butcher shop with excellent plant-based proteins and cheeses.

Food waste is another issue that is responsible for roughly eight percent of global greenhouse gas emissions. The FDA offers great tips to avoid food waste at their website **fda.gov/food/consumers/tips-reduce-food-waste.** Tips include making a shopping list and sticking to it, asking for to-go containers for leftovers when eating out, and selecting imperfect produce which might otherwise go to waste. Making soups, rice stir-fries, and frittatas out of wilted vegetables is not only delicious, but saves food from going to waste. Reducing food waste not only helps protect the climate, and saves land and water, it saves you money.

Slowly changing to a plant-based diet is great for the planet and your health. Give it a try with a meatless Monday to start.

After the *Alameda Sun* went out of business, I submitted a couple of my columns to the Alameda Post, an online newspaper, and was accepted as a monthly columnist. I have received a lot of positive feedback about my monthly climate protection column. My grammar is not perfect. So, I downloaded the free version of Grammarly which provides suggested grammar corrections—a very handy tool that works great!

I also reached out to others on Facebook. I did a series where I took pictures of easy-to-make, delicious vegetarian dishes and included the recipes. You could do green tips this way as well. I do not have a Twitter or other social media accounts but that is another avenue for influencing other people.

Chapter 4

Be Bold and Say Yes!

As my name got out there with my checklist and newspaper contributions, new opportunities arose to make a difference to protect the climate. The thing to keep in mind is even though you may have not done something before does not mean you cannot do it. Be bold and say yes! Get help if needed.

For example, an Alameda Council member ran into me at a community party and asked if I would present my checklist at his upcoming town hall meeting. Even though I had done nothing like that before, I immediately said yes! On the day of the presentation, I gathered a few props like a reusable coffee mug, reusable water bottle, reusable bag, cloth napkins, and so forth, and spoke for 10-15 minutes on what individuals can do to protect the climate. I was nervous, but the reception I got was terrific. Everyone got a copy of the checklist.

That was the start of my climate protection talks. I got more talks lined up by reaching out to the non-profits I listed in the resources portion of the checklist. I also did public talks by locating a venue (a coffee shop or the public library) and issuing a press release in the local paper. A press release can be easy to do. All I had to do was include a title, date, time, location, and description, sending it labeled press release to the editor of the newspaper to invite the public to attend.

One time I received an email from Sri Subramaniam, a Climate Reality Leader trained by Al Gore's organization, the Climate Reality Project. Sri had seen my checklist at a green home open house (an energy efficient, all-electric home partially made from recycled materials), where I had left a stack. He offered to team up and give joint presentations where he would present climate change, its impacts, and solutions, and then I would present constructive actions individuals can take to protect the climate and distribute the checklist. A local magazine contacted us to do an article on our presentations. Of course, we said yes, and we were photographed and interviewed for the article. This publicity got us more presentations. Each presentation took about an hour, so it was not that big of a time investment to do these presentations.

Our local power company, Alameda Municipal Power (AMP), came to one of my talks and offered to sign up people for free home energy efficiency audits on the spot. I asked if I could do that same thing at future talks and give them the sign-ups. They happily agreed. I got dozens of families to have home energy efficiency audits in this manner.

Another day, I got a call out of the blue because I had gone through my own home energy efficiency audit with AMP. They were looking for a volunteer to do a video in their home about the energy efficiency changes they had made and discuss the audit they went through. Again, I said yes even though I had never done anything like that before. This activity was more work as I had to clean and straighten up the house thoroughly for the film crew's arrival. The next thing I knew, a film crew from Los Angeles flew in, set up lighting and sound systems in my dining room, and I was being interviewed for the video. I took them around the house, and we filmed weather stripping, energy-saving appliances, fluorescent light bulbs (LEDs were not available yet), my clothesline with sheets on the line, etc. It was a fun and rewarding experience and AMP posted the video on their website.

Just keep yourself open to new opportunities to make a difference and be ready to say yes!

Chapter 5

Partner with Others

In addition to doing things on your own, it is helpful to reach out to organizations to partner with to get even more done. I find working with other like-minded people energizes me and keeps me going strong. You can also accomplish more with a group of passionate people working together.

Once I learned that over half of our town's greenhouse gas emissions were from the transportation sector, I reached out to Alameda's bicycling and walking advocacy non-profit, Bike Walk Alameda (BWA) to see how I could volunteer. I started by doing valet bike parking at events in town. It was fun talking to other bicyclists and selling BWA memberships. I also volunteered to write articles for their newsletter. After spending so much time volunteering, I was asked to join the Board as their secretary.

As a board member, I organized bike safety classes taught by League of American Bicyclists certified instructors. It was a fair amount of work to organize one of these classes with venue selection, confirming the instructor, and promoting the event. But doing so was very rewarding as hundreds of cyclists were trained to ride their bikes safely on the streets of our town and beyond. I got a grant from my employer IBM to purchase helmets that were distributed for free to all the attendees.

Another event I organized as part of BWA was an evening of photos from around the world of bicycling. My brother Leonard has traveled the world and is an avid fan of bicycling as a mode of transportation. He is also a great photographer and has taken many pictures of bicyclists all over Asia and Europe, hauling goods piled high on bicycles and through all kinds of weather. I invited a local bike shop to share one of their outfitted bikes for hauling stuff. A young Boy Scout explained how he and his dad rode their bikes to Bay Area Rapid Transit, boarded the train with their bikes, and commuted to school and work that way. The event was an inspiring evening for attendees to get out on their bikes!

I also participated in Bike to Work Days each year, volunteering at an energizer station where I handed out coffee and snacks, and encouraged those who biked to work or wherever that day. The idea behind Bike to Work Day is to provide a supportive environment for those who may not bike frequently, to get out there and give commuting by bike a try. My friend was the head of Walk and Roll to School Day at Edison Elementary School, and I helped her each year as well, encouraging kids to bike or walk to school instead of driving.

Locating your area's bicycling and walking advocacy organization and volunteering your time is another effective way you can reduce transportation greenhouse gas emissions in your community.

Another organization I joined from its beginning is Community Action for a Sustainable Alameda (CASA). In February 2008, the Alameda City Council adopted the Local Action Plan for Climate Protection. It outlined a series of steps Alameda would take to reduce local greenhouse gas emissions. In October 2008, community members, with the support of city staff and elected officials, formed CASA to support the Local Action Plan for Climate Protection. I joined its steering committee, which meets once a month and plans activities to reduce Alameda's greenhouse gas emissions, such as programs on electrification, waste reduction, climate education, and home weatherization. Being part of CASA helps my efforts toward protecting the climate. They helped me with ideas for my monthly column, paid for the ad space for my green tips of the week in the local newspaper, distributed my checklist at the weekly Farmers Market and annual Earth Day events, posted my checklist and my talks on their website, and arranged for me to give climate protection talks in town.

One CASA event we planned was a faith-based outreach program to encourage religious organizations to take steps to reduce their building's greenhouse gas emissions and to educate their congregations on how to reduce theirs as well. We booked a venue in the library and invited every faith-based organization in town to participate. We selected a few of them who had started climate protection initiatives at their places of worship to share what actions they were taking and invited Interfaith Power and Light (an organization that engages faith-based groups to take action to protect the climate) to present as well. I did my climate protection talk on what actions people can take to reduce their carbon footprints. Thirty faith-based leaders showed up! We gave them stacks of my checklist to distribute at their places of worship, and a few of them invited me to present my climate protection talk. It was not that much work to plan this event for a small group of people, and we had a large reach into the community. We held a second event again in 2023, focusing on the electrification of gas appliances.

A cool group within CASA calls themselves "trash talkers." The trash talkers realized that, although the public had good intentions, they did not know what trash went into which bins and often sorted incorrectly. The trash talkers addressed this issue by standing next to recycling, composting, and garbage bins at public events with their pickers, advising the public how to properly sort their trash. This technique helped minimize trash going to the landfill, kept the recycling and composting bins with only the appropriate materials, and trained the public so they would sort better at home.

I was at an event at Rockwall Winery, and one of the owners and her daughter the winemaker were distraught because people were not sorting properly even though they had provided separate bins with signage. Recycling items were going into the compost bin, compost items were going into the recycling bin, and everything was going into the landfill bin. I told them about CASA trash talkers and they were all in! I volunteered with the trash talkers, and we limited the trash to less than one garbage sack for an event with hundreds of people. Rockwall was thrilled.

Alameda Municipal Power (AMP) turned out to be a great partner. In addition to collecting signups for energy efficiency audits, they started supplying me with compact fluorescent lightbulbs (CFLs) to provide to attendees of my climate protection talks. I reached out to AMP to see if they would be willing to give away CFLs to Food Bank clients. After some discussion, they said yes! I reached out to the Food Bank; they supported the idea and brought them copies of my checklist to hand out with the CFLs. It was a big hit with the Food Bank clients and a very rewarding experience. You could try something similar in your town.

A friend on the Board of the Golden Gate Bird Alliance (GGBA) invited me to get involved in their project to incorporate protecting the climate in their organization's activities. Again, I said yes. The GGBA climate committee decided, among other things, to hire a sustainability manager. We decided the focus of the sustainability manager should be two-fold: habitat restoration and reducing transportation greenhouse gas emissions of our GGBA members' travel to view birds. We also started blogging about birding and climate protection. I wrote a blog about a day in the life of a climate-friendly birder that was well received.

I am also a member of the Rotary Club of Alameda. One of Rotary's seven areas of focus is protecting the environment. We have speakers each week at our meetings, so I volunteered to present my climate protection talk, which got an enthusiastic response. I also arranged to have other climate-related speakers, including a representative from AMP to present electrification rebates, the City of Alameda

Sustainability Manager to present an update to the Climate Action and Resiliency Plan, and the Rotary Plant-Based Diet Group to present the benefits of a plant-based diet. I also got a Rotary grant to give away coastal live oak saplings to the public at the city's Earth Day celebration, as well as to plant sixteen trees in a park in town. I partnered with a local organization called 100K Trees for Humanity that provided the trees, wheelbarrows, buckets, and training on how to plant them, and got the City's Recreation and Parks Department to select the park to do the planting, and provide the mulch, soil, and stakes. Rotary provided the funding for the trees and the volunteers to do the planting. It took some coordination, but was not that much work to organize. We do this once or twice a year now with my Rotary Club and everyone enjoys participating. I also wrote an article for the Rotary District newsletter encouraging other Rotary Clubs to do tree planting and created a presentation on the tree planting process to share that knowledge.

Rotary tree sapling giveaway at Alameda's Spring Shindig

Rotary tree planting in an Alameda Park

Another action I took with my Rotary Club was to get involved with our International Committee. I uncovered an international Rotary Program in Kenya where Rotarians were funding solar-powered water sanitation kits and convinced our International Committee to contribute one thousand dollars. Doing so protected the climate in several ways. Instead of cutting down trees for firewood to burn to boil water, families can boil water with the power of the sun. More trees remain to soak up carbon dioxide and emissions from the fires to boil the water are eliminated. Also, girls can spend more time in school rather than gathering firewood.

I encourage you to find partners in climate protection in your town and team up with them for greater action and moral support, and you can look up your local chapter of 350.org at 350.org. As their website says, 350.org is an international movement of ordinary people working to end the age of fossil fuels and build a world of community-led renewable energy for all. They are named 350.org because their goal is to reduce carbon dioxide concentration in the atmosphere to 350 parts per million—the limit scientists indicate will prevent the worst consequences of climate change. One recent meeting I attended had the attendees send emails to our representatives at the state level to support a bill that would reduce methane emissions and to support a bill for electric school buses in California. They also organized different events, like bike rides and hikes, to highlight the climate crisis.

Another worthwhile climate protection organization available globally is Citizen's Climate Lobby. You could join your local chapter by visiting their website at **citizensclimatelobby.org/**. They work to enact policy changes by lobbying their government representatives. I am a contributor to their Letter to the Editor Committee for the San Francisco Bay Area, encouraging and supporting members to write letters to the editor.

A United States-based group I recently joined is called Climate Change Makers. They meet for an hour once a week (lots of days and times to choose from), and act on a particular climate-related topic. They have a fantastic website that guides you through acting, climatechangemakers. org. For example, we recently contacted local and state representatives to apply for federal grants for solar panels for low-income disadvantaged populations.

Another great United States-based group I joined is called Elders Climate Action, which was set up to organize seniors to protect the climate in the United States. They have an excellent newsletter with specific policy actions to take on the spot, a monthly education call, and workshops on topics such as electrification. They also write actions for the Climate Action Now application described in the next chapter. They have several local geographically-based chapters and are adding more all the time.

The Sierra Club is also focused on climate protection, and their newsletter provides specific actions to encourage elected officials to focus on greenhouse gas emission reduction. They have committees that work on climate protection you can join. Or you could start your own climate protection organization with like-minded people in your community.

Chapter 6

Incorporate Climate Protection Action into Your Everyday Life

You can build climate protection action into your life as well while you are out and about participating in events in town. For example, I was at a popcorn-selling fundraising meeting at the Boy Scouts Alameda Council, and I noticed they only had a single garbage can for trash inside the room where we were having dinner before the meeting. I stepped outside the building and there were the three bins for recycling, composting, and garbage. So, I just stood up and announced that the bins were outside and suggested we sort our trash outside. Everyone was enthusiastic about it and the trash got sorted that evening. My simple act, which only took a minute or two, started a nice conversation about how to sort properly and one parent said she had read a very useful column in the newspaper about how to sort trash that she posted on her refrigerator. That was my column!

At a Boy Scout Troop awards dinner potluck, I set up three bins for sorting our trash and at the end of the meal, stood by the bins and helped people sort their trash. The scouts found the proper bins outside and emptied the trash properly. The only items that did not go into the recycling or composting bin were some plastic wrap and utensils. One parent took home the plastic utensils to wash for the next gathering. This effort took no extra time. I was at the potluck, anyway. When I arrived at the next year's awards dinner, another parent had already set up the three bins!

I noticed there were no composting or recycling bins at my son's Little League fields. I contacted the head of Little League to politely ask if we could add composting and recycling to the facilities and he put me in touch with the volunteer who oversaw garbage disposal—Tony. I gave Tony a call and explained what I wanted to do and that I was willing to help. His interest was in vector control rather than the environment, but he said he was willing to add the other bins if I could teach him about sorting and make signs of the trash items from the snack bar indicating what goes into each bin. I readily agreed, and he contacted our local waste company to get additional bins. I explained that the bins needed to be clustered in threes—composting, recycling, and garbage—or they all just become garbage bins. Tony got it, and that season we had sorting at the Little League fields. It was a pleasure seeing kids and parents stop to look at the colorful signs I made before disposing of their items at the fields. Then Tony took it to the next level and spoke to the snack bar about reducing items that would end up in the trash. So, Tony went from someone not really interested in the environment to stopping trash at its source! And all it took was a couple of phone calls and making some signs for each bin.

You may find that people around you start acting to protect the climate as well. My friend and neighbor Joe went to his son's baseball game in another town and was finishing his Starbucks coffee. The person at the gate said he had to empty his cup into a plain Styrofoam cup before entering the stands. He politely asked to see someone in charge and then explained that this was a waste of cups and that the Styrofoam one would not decompose. Joe ended up finishing his cup of coffee outside the stands, but the man in charge came over and sat down next to him, letting him know he had spoken to the snack bar and they would no longer order Styrofoam cups. A small victory that took very little of Joe's time. The key was being calm and polite, with the goal to turn the other person's view around.

My daughter was in the theatre at high school, which meant a lot of rehearsals where the kids needed water. Parents were talking about buying cases of bottled water. I offered to bring a large igloo water dispenser to fill with tap water instead, which was much more environmentally friendly and saved the parents' money, too. This act again was not much effort, especially since where I live has great tasting water.

Taking additional climate protection actions by building them into your everyday life is not that time-consuming depending on the action, or that difficult once you get started.

Chapter 7

Political Action Is Crucial as Well

It is not enough to have individuals reduce their carbon footprints on their own. Implementing governmental policies can have a broad and powerful impact as well. So how can an individual influence government policy to protect the climate?

Writing letters to the editor helps as local politicians read those letters to keep in touch with their constituencies.

I also research politicians at all levels of government to see their stance on protecting the climate. Then I vote for those who will protect the climate. If you are not registered to vote, go to vote.org to register. I also contribute to and volunteer for climate-friendly politicians. This does take more time, but you can do as much or as little as you want, from texting, phone banking, or postcard writing, to walking a precinct. I have done all those items for candidates I want to support.

Then once in office, it is important to communicate to your elected officials at all levels that they take action to protect the climate. The order of effectiveness of methods to reach your representatives is meeting them in person, calling their office, emailing, or writing them a letter, and signing a petition. In the US, there is an app for emailing or calling your representatives called Climate Action Now. I downloaded it on my iPhone and send ten pre-written letters a day first thing each morning. That only takes me about three minutes. The letters go to state and federal representatives based on your zip code. The app also has pre-written letters you can send to corporations encouraging them to take action to reduce their greenhouse gas emissions, as well as some educational types of actions. The app features having a tree planted based on the points you accumulate by taking actions in the app. So far, they have planted 217 trees on my behalf. The app is free, but you can also make a onetime or monthly donation to support their efforts. The folks behind Climate Action Now also occasionally hold Action Now parties on Zoom. They present one topic, and then everyone on the call does actions centered around that topic.

The Climate Action Now app does not send letters to local politicians like your city council and mayor. I found their email addresses on the city's website. I also speak up at city council meetings when environmental topics come up like the city's urban forest plan, proposed bike lanes, etc. It helps to be a member of local organizations to be alerted to critical items on the city council agenda. The first time I spoke I was quite nervous, but I just wrote out my comments in advance, made sure they fit into the time allotted for public comment, and read them to the council. When speaking at the council is impossible, I submit a public comment by email through the city's website. Just be sure to specify the agenda item you are writing about so it can be entered into the record. Councils can be swayed by public comment. For example, we got a protected bike lane in town after many bicyclists—including me—spoke in favor of it, outnumbering those against it.

Cities often have public workshops to garner public input on various topics. Over the years, I have participated in climate action and resiliency plan development workshops, new park workshops, new bike lanes workshops, and the city's urban forest (trees planted in our city) plan workshop. The typical workshop consists of the city presenting its tentative plans and soliciting public feedback. By providing feedback, you can change your community!

Chapter 8

Taking It to the Next Level—
Climate Reality Leader

If you have got more time, you can train to become a Climate Reality Leader with the Climate Reality Project. I took this three-day training in Minneapolis after retiring, joining tens of thousands of others who give climate reality talks in their communities. The mission of the Climate Reality Project, founded by former US Vice President Al Gore is: "to catalyze a global solution to the climate crisis by making urgent action a necessity across every sector of society. We recruit, train, and mobilize people of all walks of life to work for just climate solutions that speed energy transition worldwide and open the door to a better tomorrow for us all."

They train Climate Reality Leaders to give presentations with a very comprehensive slide deck. The presentation covers a bit about the science behind climate change, facts to show the temperature of the earth is increasing, the impacts of climate change, and the solutions we have available now to address it. I also add what individuals can do about climate protection, reviewing the checklist with them.

To find audiences when I got home, I started by giving presentations to a group of friends. Then they invited me to present to their friends and groups, like a local business association and the League of Women Voters. It spread from there. When the pandemic hit, I switched to giving presentations via Zoom. The most fun one I gave was at a winery.

The owner had asked what I was doing in retirement and when I explained I was giving climate protection presentations, she told me she shared my passion and invited me to present at her winery. I included a few charts about how climate change is affecting the wine business. For example, if warming continues unabated, Napa Valley will no longer be able to grow its famous Cabernet Sauvignon grapes. I have also given presentations to businesses interested in being more environmentally conscious. I joined the San Francisco Bay chapter of the Climate Reality Project as well. They occasionally send out pleas for presentations from requests they receive through their website.

To learn when the next Climate Reality Project training occurs, visit their website at **climaterealityproject.org.**

Author Joyce Mercado giving a Climate Reality Presentation at a friend's home

Chapter 9

Have Faith

It is easy to get discouraged and wonder if you are truly making a difference during your journey toward becoming a climate protection activist. But have a little faith. You *will* get positive feedback about what you are doing.

I hit a low point when bicycling to the grocery store and I was passed by a Hummer with one person in it. Was I truly making a difference? But after I did my shopping and was loading up my bike, another shopper stopped by and asked me about how I fit so much on my bike. I showed her the basket and paniers which attach to the bike rack in the back, and my bungee cord for strapping larger items to the rack. She was very interested and said she was going to modify her bike to do the same.

Another time I was at a neighbor's party and a Little League mom came running over when she saw me. I had given her the checklist at a game, and she was excited to let me know how much progress she had made on the checklist.

After my letter to the editor on putting up a clothesline, a Boy Scout dad let me know he had got one that day and put it up to dry beach towels and was already saving on his gas bill.

There was the mom at the Boy Scout popcorn-selling meeting who told me she put up my trash sorting article on her refrigerator.

When I meet someone new and they hear my name, they often say they love my articles in the newspaper.

Once at a Weight Watchers meeting, they asked the group who inspires you. Another attendee surprised me by saying I inspired her by riding my bike to the meetings each week. She said, "I cannot ride a bike, but I can walk!" And she started walking to meetings instead of driving. Years later, I ran into her, and she was still walking to places in town instead of driving. So, you never know who you are inspiring.

So have a little faith. Your actions are influencing others' behavior. You are making a difference! Once you are done reading this book, please give it to a like-minded friend to impact another community. Together we can change people's habits to become more climate-friendly—one community at a time.

About the Author

Joyce Mercado has been an effective climate activist in her community of Alameda for the last twenty years even while holding a full-time job and raising two children. She is a monthly climate protection columnist for the local Alameda press, a trained Climate Reality Project Leader who has given dozens of presentations on climate protection, and has been recognized by the City of Alameda for her climate protection work.

The climate emergency prompted her to write this book when she retired so others could learn from her experience to become climate leaders in their communities. Joyce is a proud Rotarian having served as Community Service Chair and President of the Rotary Club of Alameda. Joyce has a Bachelor of Science in Physics from California Polytechnic State University, San Luis Obispo. She went to graduate school at Princeton University to study plasma (electrically charged gas) physics for two years. Joyce decided plasma physics was not for her, attended her first on-campus interview, and got a job at IBM for the next 35 years. Joyce lives in Alameda, California with her husband David.